FLORA OF TROPICAL EAST AFRICA

PLANTAGINACEAE

B. VERDCOURT

Annual or perennial, terrestrial or, in one small genus, aquatic herbs. Leaves mostly spirally arranged and all radical, less often cauline, alternate or opposite, simple, often sheathing at the base, sometimes reduced, exstipulate. Flowers small, hermaphrodite, or in one small genus unisexual, regular, sessile in the bract-axils, forming axillary spikes or capitate racemes. Calyx herbaceous, 4-lobed, the lobes ± free or sometimes the lower pair ± united. Corolla gamopetalous, scarious, 3–4-lobed, the lobes imbricate. Stamens (1–2–)4, inserted in the corolla-tube and alternating with the lobes, rarely hypogynous; filaments usually long; anthers versatile, 2-thecous, the thecae opening lengthwise. Ovary superior, 1–4-locular; style long, simple, often exserted; ovules 1–∞ in each locule, axile or basal. Fruit a membranous circumscissile capsule or a 1-seeded bony nut. Seeds peltately attached, often sticky when wet; embryo straight, rarely curved, placed in the middle of the fleshy endosperm.

A rather large family of 3 genera, one monotypic, another with only 3 species and the other with over 250 species; widely distributed in both hemispheres, mostly in temperate regions. Only one genus occurs in the Flora area.

PLANTAGO

L., Sp. Pl.: 112 (1753) & Gen. Pl., ed. 5: 52 (1754); Pilg. in E.P. IV. 269: 1–466 (1937)

Annual or perennial, terrestrial, caulescent or mostly acaulescent herbs or suffruticose herbs. Leaves usually all radical but sometimes, in the species with developed stems, cauline. Flowers nearly always hermaphrodite, 4-merous, in heads or cylindrical spikes. Stamens inserted in the corolla-tube. Ovary 2–4-locular; ovules 1–many in each locule. Capsule circumscissile. Seeds 1–many in each locule, usually ± boat-shaped with ventral hilum; albumen fleshy; embryo straight or curved; radicle inferior.

A cosmopolitan genus of some 265 species, several of them pernicious weeds; 6 occur in the Flora area but only 3 are native.

Species 1–5 belong to subgen. *Plantago* and are distributed in 3 sections: sect. *Polyneuron* Decne. (sp. 1), sect. *Palaeopsyllium* Pilg. (sp. 2–4) and sect. *Arnoglossum* Decne. (sp. 5); species 6 belongs to subgen. *Psyllium* (Juss.) Harms sect. *Psyllium*.

P. ovata Forsk. has been cultivated at Kabete, Kenya (Oct. 1934, *Hudson* in *A.D.* 2936!). It will key to near *P. lanceolata* in the key below but is not closely related, differing in having 4 free sepals and woolly hairs on the leaves, which are linear or lanceolate.

Leaves all radical, alternate and spirally arranged:
  Leaves not deeply lobed, entire, sinuate-dentate or remotely lobulate:
    Leaves elliptic, oblong, ovate or almost round; ovary with each locule containing 1–many ovules:
      Spikes dense; leaf-blade gradually cuneate into the petiole; ovules 3–many per locule:

Capsule globose, 2–3 mm. in diameter (in the
Flora area); fruiting inflorescences slender,
mostly 5–8 mm. wide (introduced weed)   .   1. *P. major*
Capsule narrowly ovoid, 5 mm. long, 3 mm. wide;
fruiting inflorescences distinctly more robust,
mostly about 1 cm. wide (native plant)   .   3. *P. africana*
Spikes lax, particularly below; leaf-blade passing
more abruptly into the petiole; ovule 1 in each
locule   .   .   .   .   4. *P. fischeri*
Leaves lanceolate; ovary with only 1 ovule in each
locule   .   .   .   .   5. *P. lanceolata*
Leaves very distinctly palmately lobed; ovules 2–5 in
each locule   .   .   .   .   .   .   2. *P. palmata*
Leaves cauline, opposite   .   .   .   .   .   6. *P. afra*

1. **P. major** *L.*, Sp. Pl.: 112 (1753); Bak. in F.T.A. 5: 503 (1900); Pilg. in
E.P. IV. 269: 41 (1937); Verdc. in Journ. E. Afr. Nat. Hist. Soc. 24 (108): 60
(1964); Sagar & Harper in Journ. Ecol. 52: 189–205 (1964).   Type: perhaps
Sweden, specimen 144.1* (LINN, lecto.!)

An extremely variable, usually robust, glabrous or pubescent herb, with
short stout erect stem and numerous adventitious roots.   Leaves borne in a
rosette, spirally arranged; blade ovate, elliptic or rarely rounded, (1–)10–
20(–30) cm. long, (0·7–)5–9·5(–17) cm. wide, rounded at the apex, entire,
sinuate or irregularly toothed, 3–9-nerved at the base, abruptly narrowed at
the base into a petiole usually ± equalling the blade, 1–20 cm. long.   Inflorescences spicate, the spikes nearly always cylindric, almost invariably
simple, (0·5–)10–15(–70) cm. long; peduncles (1–)7–15(–40) cm. long; bracts
1–2·5 mm. long, acute, brownish with a brown keel.   Sepals broadly elliptic
to rounded, 1·5–2 mm. long and wide, keeled.   Corolla greenish- or yellowish-
white, 2–4 mm. long; lobes elliptic-ovate, ovate or narrowly triangular,
1–1·25 mm. long, obtuse or acute.   Anthers at first lilac, later whitish or
yellowish.   Capsule globose or subconic, 2–3 mm. long, 3–34-seeded.   Seeds
brown to dark brown or olivaceous with lighter belt on back, ellipsoidal or
angular, 1–1·7 mm. long, 0·8 mm. wide, the ventral side more convex than
the dorsal side; hilum lighter.   Fig. 1/1, 2.

KENYA. Nairobi, banks of Nairobi R., 21 Feb. 1960, *G. R. Williams* 732! & Aug. 1960,
Hussain in *E.A.H.* H221/60! & 26 Jan. 1961, *Verdcourt* 3046!
DISTR. **K**4; S. Tomé, Congo, Ethiopia, Rhodesia, Angola and South Africa; ranging
naturally throughout Europe and northern and central Asia, but now naturalized
throughout most of the world
HAB. Grassy river-banks; 1665 m.

2. **P. palmata** *Hook. f.* in J.L.S. 6: 19 (1861) & 7: 213 (1864); Engl., P.O.A.
C: 374 (1895); Bak. in F.T.A. 5: 504 (1900); Thonner, Blütenpfl. Afr., t. 143
(1908). Pilg. in E.P. IV. 269: 77 (1937); Hepper in F.W.T.A., ed. 2, 2: 306,
fig. 271 (1963); Verdc. in Journ. E. Afr. Nat. Hist. Soc. 24 (108): 60 (1964).
Type: Fernando Po, Clarence Peak, *Mann* 611 (K, holo.!)

Perennial, glabrescent or ± hairy herb with short, stout rhizome and
numerous roots.   Leaves borne in a rosette, spirally arranged; blade ovate or
rounded, distinctly palmatilobed, 5–12 cm. long, 3–13 cm. wide, acute or
rounded at the apex, cordate, hairy above and on the nerves beneath, later
often glabrescent or glabrous, 3–7-nerved at the base; petiole mostly long,
4–29 cm. long, dilated at the base, sparsely to densely hairy.   Inflorescences
spicate, the spikes mostly dense, but occasionally lax at the base, 1–17 cm.

* Reference is to Hort. Cliff., but specimens not suitable.

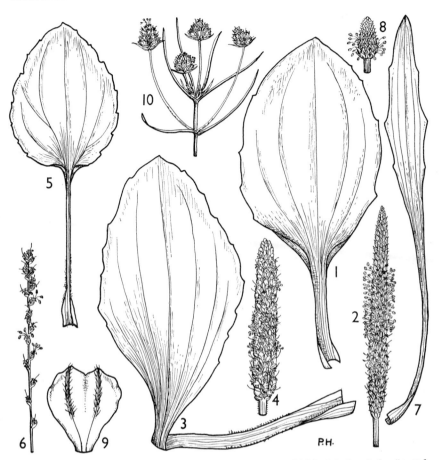

FIG. 1. *PLANTAGO MAJOR*—**1**, leaf, × ⅔; **2**, spike, × ⅔. *P. AFRICANA*—**3**, leaf, × ⅔; **4**, spike, × ⅔. *P. FISCHERI*—**5**, leaf, × ⅔; **6**, spike, × ⅔. *P. LANCEOLATA*—**7**, leaf, × ⅔; **8**, spike, × ⅔; **9**, lower pair of sepals, × 8. *P. AFRA*—**10**, flowering shoot, × ⅔.

long; peduncles 5–30 cm. long, sulcate, glabrescent to hairy, usually densely hairy just beneath the spike; bracts broadly elliptic, 2·5–3·5 mm. long, 2 mm. wide, keeled with green. Sepals oblong, rounded elliptic or obovate, 2·5–3 mm. long, 1·5–2·5 mm. wide, obliquely keeled. Corolla white or greenish-white, ± 3 mm. long; lobes ovate or oblong-ovate, 1·3–1·5 mm. long, 0·5–0·8 mm. wide, acute, keeled with green, sometimes tinged purple. Anthers yellowish or cream-coloured. Capsule ellipsoidal or subglobose, 3–4 mm. long, 1·5–2·8 mm. wide. Seeds opaque or slightly shining, brown, reddish-brown or olivaceous, ellipsoid, 2·5 mm. long, 1·5 mm. wide, the hilar side slightly concave or ± plane. Fig. 2, p. 4.

UGANDA. Karamoja District: Mt. Kadam [Debasien], Mareyo, Jan. 1936, *Eggeling* 2767!; Kigezi District: Kachwekano Farm, May 1949, *Purseglove* 2819!; Mbale District: Bugishu, Budadiri, Jan. 1932, *Chandler* 456!

KENYA. W. Suk District: Kapenguria, 6 May 1953, *Padwa* 67!; Kiambu District: Bamboo Forest, Kinale, 2 Sept. 1951, *Verdcourt* 600!; Masai District: 48 km. from Olokurto, edge of Olenguruone Settlement, 16 May 1961, *Glover et al.* 1147!

TANGANYIKA. Arusha District: E. slope of Mt. Meru, Jekuhama R., 21 Mar. 1966, *Greenway & Kanuri* 12455!; Kilimanjaro, Bismarck Hill to Marangu, 1 Mar. 1934, *Greenway* 3884!; Rungwe Mission, Jan. 1954, *Semsei* 1564!

FIG. 2. *PLANTAGO PALMATA*—**1**, habit, × ⅔; **2**, flower, × 8; **3**, anther, × 16; **4**, young fruit, × 8; **5**, seed, × 8. All from *Verdcourt* 600.

DISTR. U1–4; **K**2–6; **T**2, 7; Cameroun, Fernando Po, E. Congo and Rwanda, Ethiopia and Rhodesia (Inyanga)

HAB. Openings and disturbed places in montane forest, particularly bamboo and *Juniperus*, often by streams or on boggy ground, less often in thicket or in plantations; 1170–3150(–? 3300) m.

SYN. *P. kerstenii* Aschers. in Sitzb. Ges. naturf. Fr. 20 Oct. 1868: 23 (1868). Type: Tanganyika, Kilimanjaro, *Kersten* (B, holo. †, K, fragment !)
P. palmata Hook. f. var. *kerstenii* (Aschers.) Aschers. in Sitzb. Ges. naturf. Fr. 19 Mar. 1872: 38 (1872) & in von der Decken, Reisen 3(3), Bot.: 74, t. 4 (1879)
[*P. fischeri* sensu Engl. in E.J. 19, Beibl. 47: 48 (1894); P.O.A. C: 374 (1895); Bak. in F.T.A. 5: 504 (1900), quoad *Fischer 512* (B †), *non* Engl.]

3. **P. africana** *Verdc.* in K.B. 23: 507, fig. 1 (1969). Type: Kenya, Kericho District, SW. Mau Forest, *Maas Geesteranus 5778* (K, holo. !, L, iso.)

Perennial herb with stout probably horizontal rhizome at least 4·5 cm. long, 1–1·5 cm. thick. Leaves borne in a rosette, spirally arranged; blade elliptic or oblong-elliptic, 5–13 cm. long, 2·8–7 cm. wide, obtuse at the apex, gradually narrowed at the base into the petiole, ± entire or with several remote lobules on the margin, glabrescent above, pilose with crisped hairs beneath particularly on the 5–7 nerves; petiole distinct from the blade, 2–14 cm. long, usually pilose, dilated at the base, with or without cottony hairs. Inflorescences spicate, the spikes dense, 3–8 cm. long; peduncles 9–26 cm. long, densely to sparsely pilose; bracts elliptic, 2·2–3·5 mm. long, 1·2–2 mm. wide, keeled. Sepals elliptic, 2·5–3 mm. long, 1·8–2·2 mm. wide, keeled. Corolla whitish or brownish, 3 mm. long; lobes rounded-ovate, 0·7–1 mm. long, 0·6–0·7 mm. wide. Capsule narrowly ovoid, 5 mm. long, 3 mm. wide, 4–6-seeded. Immature seeds pale, semi-ellipsoid, 2·2 mm. long, 1·3 mm. wide, flattened on hilar side. Fig. 1/3, 4, p. 3.

KENYA. Kericho District: SW. Mau Forest Reserve, Camp 8, along the R.It are, 14 Aug. 1949, *Maas Geesteranus 5778*!; Masai District: Narok area. Shabal Taragwa, about 4·8 km. from Ol Pusimoru Sawmill, 20 May 1961, *Glover et al. 1413*!

DISTR. **K**5, 6; Ethiopia

HAB. Open glades and river-banks in upland evergreen forest; 1920–2550 m.

NOTE. *Whittal 39*!, from a swamp at 2400 m. in the SW. Mau Forest and consisting of a single young inflorescence and half a leaf, closely resembles *P. palmata* save that the leaf-margin is merely undulate and its base cuneate; it is in some ways intermediate with *P. africana* and it is very probable that hybrids occur.

4. **P. fischeri** *Engl.* in E.J. 19, Beibl. 47: 48 (1894) & P.O.A. C: 374 (1895); Bak. in F.T.A. 5: 504 (1900); Pilg. in E.P. IV. 269: 78, fig. 11 (1937); Verdc. in Journ. E. Afr. Nat. Hist. Soc. 24 (108): 61 (1964). Type: Tanganyika, Kilimanjaro, Mawenzi, *Volkens 948* (B, lecto. †)

Perennial herb with long stout rhizome 6–36 cm. long, covered with densely cottony-hairy remains of petiole-bases. Leaves borne in a rosette, spirally arranged; blade ovate or rounded-ovate, 4–14 cm. long, 2·5–11 cm. wide, rounded or subacute at the apex, rounded to slightly subcordate at the base, almost entire to distinctly undulate-dentate or -crenulate or less often very shallowly crenate-lobed, glabrous or almost so, 6–7-nerved; petiole abruptly distinct from the blade, 2–16 cm. long, dilated at the base, rather sparsely to quite densely covered, particularly at the base, with long, very pale brown, cottony hairs. Inflorescences laxly spicate, with the flowers well separated particularly at the base, (1–)4·5–20 cm. long; peduncles 10–40 cm. long, basally with the same cottony indumentum as the petioles; bracts oblong or ovate, 2·2–3 mm. long, 1·2 mm. wide, keeled. Sepals narrowly elliptic to elliptic-ovate, 2·5–3 mm. long, 1·5–2 mm. wide, keeled. Corolla whitish or greenish, 2·5–3 mm. long; lobes narrowly ovate, 1·3–1·5 mm. long, 0·8 mm. wide. Immature capsule ellipsoid, 3–3·5 mm. long, 1·6 mm. wide,

1-seeded. Almost ripe seeds semi-ellipsoid, 2·5 mm. long, 1·5 mm. wide, slightly concave near the hilum. Fig. 1/5, 6, p. 3.

TANGANYIKA. Arusha District: Mt. Meru, 3 Oct. 1959, *Carmichael* 727! & Mt. Meru, Crater, 5 Oct. 1932, *B. D. Burtt* 4163!; Kilimanjaro, above Moshi, Muë stream, 8 Oct. 1901, *Uhlig* 135! & SE. Kilimanjaro, 7 Apr. 1934, *Schlieben* 4901!
DISTR. **T2**; not known elsewhere
HAB. Montane forest, also moist sandy soil of water-course on lava flows; 2460–3300 m.

SYN. *P. fischeri* Engl. forma *supina* Pilg. in E.P. IV. 269 : 78 (1937). Type: Tanganyika, Kilimanjaro, Lume R., Kitorovi stream above Mkuu, *Volkens* 1901 (B, holo., BM, K, iso.!)

NOTE. Pilger states that *Fischer* 512 cited by Engler as a syntype is in fact *P. palmata* and although he does not specifically mention the word lectotype he is obviously choosing *Volkens* 948 to fulfill that rôle and Engler's description clearly fits the species under consideration. Unfortunately all the Berlin material was destroyed in the War. As has been pointed out on the Kew covers by J. P. M. Brenan, *P. fischeri* is closely similar to the Madagascan species *P. tanalensis* Bak., but that has very short rhizomes with numerous roots just below the leaf-rosette, distinctly hairy leaves and the long cottony hairs at the base of the petiole are much less developed; moreover the fruits are 2-seeded. More material of *P. fischeri* is required to see if its fruits are constantly 1-seeded.

5. **P. lanceolata** *L.*, Sp. Pl.: 113 (1753); Bak. in F.T.A. 5 : 503 (1900); Pilg. in E.P. IV. 269 : 313 (1937); Verdc. in Journ. E. Afr. Nat. Hist. Soc. 24 (108): 60 (1964); Sagar & Harper in Journ. Ecol. 52 : 211–218 (1964). Type: specimen of Plantago angustifolia major in Hort. Cliff. p. 36, No. 3 (BM, lecto.!)

An extremely variable glabrous, pubescent or more rarely densely pilose perennial herb from a ± erect, thick, short rhizome; stem silky hairy. Leaves borne in a rosette, spirally arranged; blade linear-lanceolate, ovate-lanceolate or spathulate, (2–)10–15(–45) cm. long, 0·7–3(–8) cm. wide, acute or acuminate at the apex, usually entire or somewhat toothed, 3–5(–7)-veined, mostly gradually narrowed into a petiole usually equalling the blade in length but occasionally much shorter or much longer. Inflorescence spicate, the spikes cylindric to globose, (0·3–)0·5–3·5(–10) cm. long; peduncles 5–120 cm. long, mostly furrowed and silky; bracts ovate-acuminate, 4–7 mm. long, dark blackish-green, with scarious points. Sepals rounded-ovate, 3–3·5 mm. long, glabrous or hairy on the margins or keels, the lower pair joined to form an obovate, ± bilobed 2-keeled organ. Corolla brownish-white, 3 mm. long; lobes narrowly ovate or ovate, 2–2·5 mm. long. Anthers mostly white. Capsule ellipsoid, 3–6 mm. long, 1–2-seeded. Seeds yellow-brown to dark brown, shining, oblong-ellipsoid, 2·5–3 mm. long, 1·5 mm. wide, dorsal side convex, ventral side concave; hilum dark. Fig. 1/7–9, p. 3.

KENYA. Elgeyo District: Cherangani Hills, below Kaisungor, May 1964, *Tweedie* 2806!
TANGANYIKA. Lushoto District: Gare Mission, 14 Jan. 1941, *Greenway* 6108! & Lushoto, 11 Jan. 1941, *Greenway* 6096! & Shume Forest Reserve, Aug. 1955, *Semsei* 2331!
DISTR. **K3**; **T3**; Gabon, Ethiopia, Sudan, Malawi, Rhodesia, Botswana and South Africa; ranging naturally throughout Europe and Asia, but now naturalized throughout most of the world.
HAB. Roadsides in wooded and open grassland; 1500–2400 m.

6. **P. afra** *L.*, Sp. Pl., ed. 2 : 168 (1762); Grande in N. Giorn. Bot. Ital., n. ser., 32 : 76 (1925). Type: Malta and N. Africa*, Moris. Hist. 3 : 262, sect. 8, t. 17/4 (syn.!)

* Linnaeus' reference to N. Africa and his description probably derive from Brander's specimen from Algiers (LINN, specim. 144/31).

Annual herb 15–50 cm. tall, erect or ascending; stems with a mixture of rigid hairs and minute glandular hairs. Leaves opposite, linear to linear-lanceolate, 1·5–6 cm. long, 1–5 mm. wide, entire or with a few teeth, narrowed at both ends, glabrescent or with short and long hairs mixed, sometimes villous near the base. Inflorescences borne in upper leaf-axils, the spikes small, (0·5–)0·8–1·5 cm. long; peduncles 1–5 cm. long; bracts ovate or narrowly ovate, 3·5–5 mm. long, drawn out into a long acumen at the apex, altogether ± 8 mm. long. Sepals 3–4 mm. long, hairy and glandular, the posticous pair narrowly ovate to broadly oblanceolate. Corolla cream to yellow, ± 4 mm. long; lobes ovate, 2 mm. long. Capsule ellipsoid or rounded-ellipsoid, 3–3·5 mm. long. Seeds reddish-brown, shining, oblong-ellipsoid, 2–3·25(–4) mm. long, convex on dorsal surface, concave-sulcate on hilar surface. Fig. 1/10, p. 3.

SYN. [*P. psyllium* sensu L., Sp. Pl., ed. 2: 167 (1762) et auctt. mult., e.g. Pilg. in E.P. IV. 269: 422, fig. 43 A (1937), *non* L., Sp. Pl.: 115 (1753)]
      P. cynops L., Sp. Pl.: 116 (1753), *non* sensu auctt. mult., *nom. ambig.* Type: plant grown in Hort. Upsal. (not found)

var. **stricta** (*Schousb.*) *Verdc.* in K.B. 23: 509 (1969). Types: Morocco, Mogodor, *Schousboe* (? C, syn.); Moris., Hist. 3, sect. 8, t. 17/2 (syn.!)

Slender plant 15–35 cm. tall, with narrow leaves 1–2 mm. wide mostly without teeth.

TANGANYIKA. Masai District: Ngorongoro Crater floor, 3 Mar. 1961, *Newbould* 5687! & Malenda, 30 Apr. 1961, *Newbould* 5838!
DISTR. T2; Morocco, Algeria, Egypt to Ethiopia, Israel, Syria and Canary Is.
HAB. Overgrazed *Pennisetum massaicum* grassland; 1500–2400 m.

SYN. *P. stricta* Schousb., Iagtt. Vextr. Marokko: 69 (1800)
      P. psyllium L. var. *stricta* (Schousb.) Maire in Jahandiez & Maire, Cat. Pl. Maroc. 3: 706 (1934); Pilg. in E.P. IV. 269: 424 (1937); Verdc. in Journ. E. Afr. Nat. Hist. Soc. 24 (108): 61 (1964)

NOTE. Linnaeus changed the application of the names *P. psyllium* and *P. cynops* in Sp. Pl., ed. 2. Unfortunately the species described above has universally been called *P. psyllium* and it is the greatest pity that this specific name cannot be conserved. What has long been known as *P. cynops* L. should be called *P. sempervirens* Crantz.

INDEX TO PLANTAGINACEAE